Ray's New Arithmetic Workbook

Series 2 - Book 3

Rudolph Moore, Ph.D.
Betty Moore, M.A.

© Mott Media, L.L.C.
Fenton, Michigan

Series 2 - Book 3

All rights reserved. No portion of this book may be reproduced without written permission from the publisher except by a reviewer who may quote brief passages in connection with a review. Write to Editor, Mott Media, L.L.C., 1130 Fenway Circle, Fenton, Michigan 48430.

About the *Classic Curriculum* Arithmetic Workbooks

The *Classic Curriculum* arthimetic workbooks are available for the first four years of arithmetic instruction. They are specifically written for use with the *Ray's Arithmetic Series* of textbooks (also published by Mott Media). The four Series One workbooks teach the very basic skills needed before beginning the textbooks. Series Two, Three, and Four workbooks require the use of *Ray's Primary Arithmetic* and later *Ray's Intellectual Arithmetic*. The *Ray's Parent/Teacher Guide* is also a valuable tool. It will be especially helpful as your child continues his math education in Ray's textbooks beyond year four.

Series Two books are for year two. There are four books in Series Two. Each workbook is designed to provide nine weeks of instruction. On days one through four of each week the child is to complete a lesson which presents some new material. Each day's work should be checked and any problems missed should be corrected by the child. On day five the child completes a review lesson and/or quiz which allows the parent to evaluate his or her child's understanding of the material to date.

The last four sheets are the final test for this workbook, answers for the final test, and answers for the daily lessons and quizes in this workbook. Please remove these four pages before giving this workbook to your child. When the child has completed all lessons in this workbook, give the final test. Make sure the score demonstrates mastery of the concepts presented before proceeding to the next workbook in the series. As always, the child should redo any problems that were answered incorrectly.

We feel that Ray's Arithmetics will provide your child with a superior knowledge of arithmetic and hope that the *Classic Curriculum Arithmetic Workbooks* will help you and your child get off to a good start in using the Ray's series of books.

Mott Media textbooks needed to complete this workbook:
    Ray's New Primary Arithmetic

Mott Media textbooks you may find helpful:
    Ray's Parent-Teacher Guide
    Key to Ray's Primary, Intellectual, and Practical Arithmetics

Procedure Checklist:

1. Until your student can read well, you must read the directions to the student.
2. Allow the student to work independently as much as possible.
3. Score the work yourself, paying close attention to letter/number formation as well as answers.
4. Ask the student to rework wrong or careless work.
5. Rescore the corrected material.
6. Be certain to praise the student for good work.
7. The student should score at least 80% before proceeding to the next workbook.

ISBN-13: 978-0-88062-238-7
ISBN-10: 0-88062-238-5

Table of Contents

Multiplication 2

How Many Times? 16

Add, Subtract, and Multiply 20

More About Groups 23

Division 26

Multiplication 6 through 9 (Review) 50

Add, Subtract, and Multiply (Review) 52

Division 2 through 5 (Review) 54

Division 6 through 9 (Review) 56

Multiplication

I can learn multiplication by six.

READ

Remember that when we do multiplication, we are working with groups. In Lesson XLIV, page 47, we are working with groups of **six** or **six groups** of some number. Read the sentences at the top of page 47. Notice that there are two sentences with the same answer. Practice these sentences using objects.

STUDY

6 times 5 are 30
6 groups of 5 are 30

5 times 6 are 30
5 groups of 6 are 30

RECITE

Write the numbers on the lines.

1. 6 times 1 are _____

2. 6 times 2 are _____

3. 6 times 3 are _____

4. 6 times 4 are _____

5. 6 times 5 are _____

6. 6 times 6 are _____

7. 6 times 7 are _____

8. 6 times 8 are _____

9. 6 times 9 are _____

10. 6 times 10 are _____

11. 1 time 6 is _____

12. 2 times 6 are _____

13. 3 times 6 are _____

14. 4 times 6 are _____

15. 5 times 6 are _____

16. 6 times 6 are _____

17. 7 times 6 are _____

18. 8 times 6 are _____

19. 9 times 6 are _____

20. 10 times 6 are _____

2 (two)

Multiplication

| I can use multiplication to work problems. |

READ

Use objects to work the problems in Lesson XLIV. Remember that we are working with groups of **six** or **six groups** of some number.

Many of these story problems are about money. We must learn to handle money correctly and wisely.

STUDY

If 1 dress can be made from 6 yards of calico, how many yards will it take to make 2 dresses.

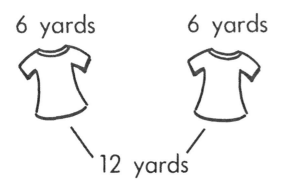

2 dresses at 6 yards each
2 times 6 are 12

RECITE
Use objects if necessary to answer all questions 1 through 9 in Lesson XLIV, page 47. Ask your teacher/tutor to ask **you** the questions.

Record your answers here.

1. _____12 yards_____

2. _____   6. _____

3. _____   7. _____

4. _____   8. _____

5. _____   9. _____

3 (three)

Multiplication

I can learn multiplication by seven.

READ

Remember that when we do multiplication, we are working with groups. In Lesson XLV, page 48, we are working with groups of **seven** or **seven groups** of some number.

Read the sentences at the top of page 48. Notice that there are two sentences with the same answer. Practice these sentences using objects.

STUDY

7 times 4 are 28
7 groups of 4 are 28

4 times 7 are 28
4 groups of 7 are 28

RECITE

Write the numbers on the lines.

1. 7 times 1 are _____

2. 7 times 2 are _____

3. 7 times 3 are _____

4. 7 times 4 are _____

5. 7 times 5 are _____

6. 7 times 6 are _____

7. 7 times 7 are _____

8. 7 times 8 are _____

9. 7 times 9 are _____

10. 7 times 10 are _____

11. 1 time 7 is _____

12. 2 times 7 are _____

13. 3 times 7 are _____

14. 4 times 7 are _____

15. 5 times 7 are _____

16. 6 times 7 are _____

17. 7 times 7 are _____

18. 8 times 7 are _____

19. 9 times 7 are _____

20. 10 times 7 are _____

Multiplication

| I can use multiplication to work problems. |

READ
Use objects to work the problems in Lesson XLV. Remember that we are working with groups of **seven** or **seven groups** of some number.

Many of these story problems are about money. We must learn to handle money correctly and wisely.

STUDY
Sarah bought 2 thimbles at 7 cents each. How much do both thimbles cost?

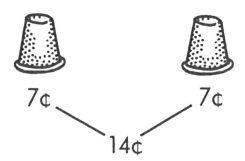

2 thimbles at 7¢ each
2 times 7 are 14

RECITE
Use objects if necessary to answer all questions 1 through 9 in Lesson XLV, page 48. Ask your teacher/tutor to ask **you** the questions.

Record your answers here.

1. _____14¢_____

2. _____

3. _____

4. _____

5. _____

6. _____

7. _____

8. _____

9. _____

5 (five)

Review I

When we multiply, we are using groups of objects. The group may have one, two, three, four, or five objects in it. We may also be talking about one, two, three, four, or five groups of some number.

The group may have six objects in it. We may be talking about six groups of some number.

six groups of two

```
*  *  *  *  *  *
*  *  *  *  *  *
```

two groups of six

```
* * *      * * *
* * *      * * *
```

The group may have seven objects in it. We may be talking about seven groups of some number.

seven groups of four

```
* *     * *     * *     * *
* *     * *     * *     * *


    * *     * *     * *
    * *     * *     * *
```

four groups of seven

```
* * *          * * *
  *              *
* * *          * * *

* * *          * * *
  *              *
* * *          * * *
```

Write the numbers on the lines.

1. 6 times 2 are _____

2. 6 times 3 are _____

3. 6 times 4 are _____

4. 6 times 5 are _____

5. 6 times 6 are _____

6. 6 times 7 are _____

7. 6 times 8 are _____

8. 6 times 9 are _____

9. 6 times 10 are _____

10. 7 times 2 are _____

11. 7 times 3 are _____

12. 7 times 4 are _____

13. 7 times 5 are _____

14. 7 times 6 are _____

15. 7 times 7 are _____

16. 7 times 8 are _____

17. 7 times 9 are _____

18. 7 times 10 are _____

19. 4 times 6 are _____

20. 5 times 6 are _____

21. 6 times 6 are _____

22. 7 times 6 are _____

23. 8 times 6 are _____

24. 9 times 6 are _____

25. 4 times 7 are _____

26. 5 times 7 are _____

27. 6 times 7 are _____

28. 7 times 7 are _____

29. 8 times 7 are _____

30. 9 times 7 are _____

Quiz I

Write the numbers on the lines.

1. 6 times 2 are _____

2. 7 times 4 are _____

3. 6 times 4 are _____

4. 6 times 3 are _____

5. 4 times 6 are _____

6. 6 times 7 are _____

7. 7 times 7 are _____

8. 8 times 6 are _____

9. 6 times 5 are _____

10. 7 times 3 are _____

11. 3 times 7 are _____

12. 9 times 6 are _____

13. 7 times 5 are _____

14. 6 times 8 are _____

15. 6 times 10 are _____

16. 3 times 9 are _____

17. 5 times 7 are _____

18. 8 times 7 are _____

19. 7 times 9 are _____

20. 7 times 10 are _____

21. Mother has 7 nickels which is _____ cents.

22. 5 boxes each having 10 pencils are _____ pencils.

23. 3 strings of beads with 6 beads each are _____ beads.

24. One shirt takes 3 yards, three shirts takes _____ yards.

25. Each house has 8 windows. 7 houses have _____ windows.

Answer five oral questions from Lessons XLIV or XLV, pages 47 or 48. These questions should be asked by a teacher/tutor.

Multiplication

I can learn multiplication by eight.

READ

Remember that when we do multiplication, we are working with groups. In Lesson XLVI, page 49, we are working with groups of **eight** or **eight groups** of some number. Read the sentences at the top of page 49. Notice that there are two sentences with the same answer. Practice these sentences using objects.

STUDY

8 times 6 are 48
8 groups of 6 are 48

6 times 8 are 48
6 groups of 8 are 48

RECITE

Write the numbers on the lines.

1. 8 times 1 are _____

2. 8 times 2 are _____

3. 8 times 3 are _____

4. 8 times 4 are _____

5. 8 times 5 are _____

6. 8 times 6 are _____

7. 8 times 7 are _____

8. 8 times 8 are _____

9. 8 times 9 are _____

10. 8 times 10 are _____

11. 1 time 8 is _____

12. 2 times 8 are _____

13. 3 times 8 are _____

14. 4 times 8 are _____

15. 5 times 8 are _____

16. 6 times 8 are _____

17. 7 times 8 are _____

18. 8 times 8 are _____

19. 9 times 8 are _____

20. 10 times 8 are _____

8 (eight)

Multiplication

| I can use multiplication to work problems. |

READ
Use objects to work the problems in Lesson XLVI. Remember that we are working with groups of **eight** or **eight groups** of some number.

Many of these story problems are about money. We must learn to handle money correctly and wisely.

STUDY
James bought 2 melons at 8 cents each. How many cents did the melons cost?

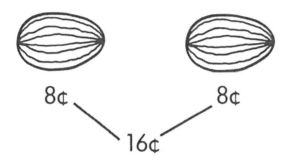

2 melons at 8¢ each
2 times 8 are 16

RECITE
Use objects if necessary to answer all questions 1 through 9 in Lesson XLVI, page 49. Ask your teacher/tutor to ask **you** the questions.

Record your answers here.

1. _____ 16¢ _____

2. _____   6. _____

3. _____   7. _____

4. _____   8. _____

5. _____   9. _____

9 (nine)

Multiplication

I can learn multiplication by nine.

READ

Remember that when we do multiplication, we are working with groups. In Lesson XLVII, page 50, we are working with groups of **nine** or **nine groups** of some number. Read the sentences at the top of page 50. Notice that there are two sentences with the same answer. Practice these sentences using objects.

STUDY

9 times 5 are 45
9 groups of 5 are 45

5 times 9 are 45
5 groups of 9 are 45

RECITE

Write the numbers on the lines.

1. 9 times 1 are _____

2. 9 times 2 are _____

3. 9 times 3 are _____

4. 9 times 4 are _____

5. 9 times 5 are _____

6. 9 times 6 are _____

7. 9 times 7 are _____

8. 9 times 8 are _____

9. 9 times 9 are _____

10. 9 times 10 are _____

11. 1 time 9 is _____

12. 2 times 9 are _____

13. 3 times 9 are _____

14. 4 times 9 are _____

15. 5 times 9 are _____

16. 6 times 9 are _____

17. 7 times 9 are _____

18. 8 times 9 are _____

19. 9 times 9 are _____

20. 10 times 9 are _____

10 (ten)

Multiplication

I can use multiplication to work problems.

READ

Use objects to work the problems in Lesson XLVII. Remember that we are working with groups of **nine** or **nine groups** of some number.

Many of these story problems are about money. We must learn to handle money correctly and wisely.

STUDY

Francis bought 2 knives at 9 cents each. How many cents did the knives cost?

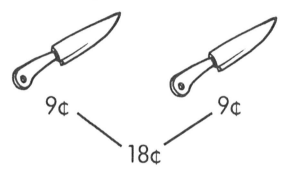

2 knives at 9¢ each
2 times 9 are 18

RECITE
Use objects if necessary to answer all questions 1 through 9 in Lesson XLVII, page 50. Ask your teacher/tutor to ask **you** the questions.

Record your answers here.

1. _____18¢_____

2. _____   6. _____

3. _____   7. _____

4. _____   8. _____

5. _____   9. _____

11 (eleven)

Review II

When we multiply, we are using groups of objects. The group may have one, two, three, four, five, six, or seven objects in it. We may also be talking about one, two, three, four, five, six, or seven groups of some number.

The group may have eight objects in it. We may be talking about eight groups of some number.

eight groups of three

```
* * *   * * *   * * *   * * *
* * *   * * *   * * *   * * *
```

three groups of eight

```
* * * *   * * * *   * * * *
* * * *   * * * *   * * * *
```

The group may have nine objects in it. We may be talking about nine groups of some number.

nine groups of two

```
* *     * *     * *     * *
            * *
* *     * *     * *     * *
```

two groups of nine

```
* * *         * * *
* * *         * * *
* * *         * * *
```

Write the numbers on the lines.

1. 8 times 2 are _____
2. 8 times 3 are _____
3. 8 times 4 are _____
4. 8 times 5 are _____
5. 8 times 6 are _____
6. 8 times 7 are _____
7. 8 times 8 are _____
8. 8 times 9 are _____
9. 8 times 10 are _____
10. 9 times 2 are _____
11. 9 times 3 are _____
12. 9 times 4 are _____
13. 9 times 5 are _____
14. 9 times 6 are _____
15. 9 times 7 are _____

16. 9 times 8 are _____
17. 9 times 9 are _____
18. 9 times 10 are _____
19. 4 times 8 are _____
20. 5 times 8 are _____
21. 6 times 8 are _____
22. 7 times 8 are _____
23. 8 times 8 are _____
24. 9 times 8 are _____
25. 4 times 9 are _____
26. 5 times 9 are _____
27. 6 times 9 are _____
28. 7 times 9 are _____
29. 8 times 9 are _____
30. 9 times 9 are _____

12 (twelve)

Quiz II

Write the numbers on the lines.

1. 8 times 2 are _____

2. 9 times 4 are _____

3. 8 times 4 are _____

4. 8 times 3 are _____

5. 4 times 8 are _____

6. 6 times 9 are _____

7. 7 times 9 are _____

8. 8 times 8 are _____

9. 8 times 5 are _____

10. 9 times 3 are _____

11. 3 times 8 are _____

12. 9 times 8 are _____

13. 9 times 5 are _____

14. 8 times 6 are _____

15. 8 times 10 are _____

16. 8 times 9 are _____

17. 5 times 9 are _____

18. 2 times 9 are _____

19. 9 times 9 are _____

20. 9 times 10 are _____

21. Father has 9 dimes. He has _____ cents.

22. 5 nests each having 8 eggs are _____ eggs.

23. One orange costs 10¢. Nine oranges cost _____ cents.

24. Jack has 4 pockets with 8 marbles in each pocket which is

_____ marbles.

25. One comb costs 8¢. Six combs cost _____ cents.

Answer five oral questions from Lessons XLVI or XLVII, pages 49 or 50. These questions should be asked by a teacher/tutor.

13 (thirteen)

Multiplication

I can learn multiplication by ten.

READ

Remember that when we do multiplication, we are working with groups. In Lesson XLVIII, page 51, we are working with groups of **ten** or **ten groups** of some number. Read the sentences at the top of page 51. Notice that there are two sentences with the same answer. Practice these sentences using objects.

STUDY

10 groups of 4

4 groups of 10

RECITE

Write the numbers on the lines.

1. 10 times 1 are _____

2. 10 times 2 are _____

3. 10 times 3 are _____

4. 10 times 4 are _____

5. 10 times 5 are _____

6. 10 times 6 are _____

7. 10 times 7 are _____

8. 10 times 8 are _____

9. 10 times 9 are _____

10. 10 times 10 are _____

11. 1 time 10 is _____

12. 2 times 10 are _____

13. 3 times 10 are _____

14. 4 times 10 are _____

15. 5 times 10 are _____

16. 6 times 10 are _____

17. 7 times 10 are _____

18. 8 times 10 are _____

19. 9 times 10 are _____

20. 10 times 10 are _____

14 (fourteen)

Multiplication

I can use multiplication to work problems.

READ

Use objects to work the problems in Lesson XLVIII. Remember that we are working with groups of **ten** or **ten groups** of some number.

Many of these story problems are about money. We must learn to handle money correctly and wisely.

STUDY

I bought 10 pencils at 2 cents each.
How much did the pencils cost?

2¢ 2¢ 2¢ 2¢ 2¢ 2¢ 2¢ 2¢ 2¢ 2¢
20¢

10 pencils at 2¢ each
10 times 2 are 20

RECITE

Use objects if necessary to answer all questions 1 through 9 in Lesson XLVIII, page 51. Ask your teacher/tutor to ask **you** the questions.

Record your answers here.

1. _____ 20¢ _____

2. _____   6. _____

3. _____   7. _____

4. _____   8. _____

5. _____   9. _____

How Many Times?

I can multiply to find how many times.

READ

The word **times** tells us to multiply. When we multiply, we are combining groups of objects. We are multiplying in Lesson XLIX, page 52.

Remember that when we do addition, we combine **single objects**. When we do multiplication, we combine **groups of objects**. Multiplication is a fast way to add.

Practice the activities on page 52 orally and in writing.* Ask someone to read the questions to you. Keep practicing until you can answer all questions without mistakes.

RECITE
Write the numbers on the lines.

1. 3 times 4 are _____

2. 2 times 9 are _____

3. 10 times 2 are _____

4. 3 times 3 are _____

5. 4 times 2 are _____

6. 5 times 4 are _____

7. 8 times 2 are _____

8. 2 times 2 are _____

9. 3 times 10 are _____

10. 5 times 6 are _____

11. 6 times 2 are _____

12. 8 times 3 are _____

13. 4 times 10 are _____

14. 7 times 3 are _____

15. 7 times 4 are _____

16. 5 times 5 are _____

17. 3 times 5 are _____

18. 6 times 3 are _____

19. 4 times 9 are _____

20. 6 times 4 are _____

21. 9 times 3 are _____

22. 4 times 8 are _____

23. 4 times 4 are _____

24. 9 times 5 are _____

25. 5 times 7 are _____

26. 10 times 8 are _____

27. 8 times 6 are _____

28. 7 times 10 are _____

29. 6 times 6 are _____

30. 8 times 8 are _____

* Use objects if necessary. Complete problems 17 through 27 on a piece of paper.

16 (sixteen)

How Many Times?

I can multiply to find how many times.

READ

Remember that the word **times** tells us to multiply. Multiplication is done by memorizing the multiplication facts we have been doing in Workbooks Two and Three. In Lesson XLIX, page 53, we are working two problems in each problem. We are multiplying twice.

STUDY

How many are 2 times 3 times 3?

Start on the right side   3 times 3 are 9
                           9 times 2 are 18

Start on the left side   2 times 3 are 6
                          6 times 3 are 18

Notice that we get the same answer either way.

RECITE
Write the numbers on the lines.*

1. 4 times 2 times 2

_____ times 2 are _____

2. 2 times 3 times 4

_____ times 4 are _____

3. 2 times 2 times 5

_____ times 5 are _____

4. 5 times 2 times 3

_____ times 3 are _____

5. 3 times 2 times 6

_____ times 6 are _____

6. 2 times 2 times 6

_____ times 6 are _____

7. 2 times 3 times 7

_____ times 7 are _____

8. 2 times 2 times 7

_____ times 7 are _____

9. 4 times 2 times 8

_____ times 8 are _____

10. 3 times 3 times 4

_____ times 4 are _____

* Use objects if necessary. Complete problems 39 through 58 on page 53 on a piece of paper.

17 (seventeen)

Review III

When we multiply, we are using groups of objects. The group may have **ten** objects in it or we may also be talking about **ten groups** of some number.

ten groups of two

```
* *   * *   * *   * *   * *

* *   * *   * *   * *   * *
```

two groups of ten

```
* * * * *     * * * * *
* * * * *     * * * * *
```

To do multiplication, we need to learn the multiplication facts. We learn these by practicing over and over.

We can also work multiplication problems that are really two problems in one. We have to multiply twice.

Write the numbers on the lines.

1. 10 times 1 are _____
2. 10 times 2 are _____
3. 10 times 3 are _____
4. 10 times 4 are _____
5. 10 times 5 are _____

6. 10 times 6 are _____
7. 10 times 7 are _____
8. 10 times 8 are _____
9. 10 times 9 are _____
10. 10 times 10 are _____

11. 9 times 7 are _____
12. 5 times 8 are _____
13. 10 times 5 are _____
14. 7 times 6 are _____
15. 6 times 9 are _____

16. 7 times 7 are _____
17. 6 times 10 are _____
18. 8 times 7 are _____
19. 8 times 9 are _____
20. 9 times 9 are _____

21. 4 times 2 times 5   _____ times 5 are _____
22. 4 times 2 times 4   _____ times 4 are _____
23. 1 times 5 times 8   _____ times 8 are _____
24. 2 times 5 times 7   _____ times 7 are _____
25. 3 times 3 times 8   _____ times 8 are _____

18 (eighteen)

Quiz III

Write the numbers on the lines.

1. 10 times 5 are _____

2. 4 times 2 are _____

3. 6 times 7 are _____

4. 10 times 9 are _____

5. 5 times 6 are _____

6. 10 times 2 are _____

7. 7 times 3 are _____

8. 9 times 4 are _____

9. 10 times 4 are _____

10. 5 times 9 are _____

11. 7 times 8 are _____

12. 4 times 7 are _____

13. 10 times 3 are _____

14. 4 times 5 are _____

15. 8 times 8 are _____

16. 10 times 6 are _____

17. 3 times 5 are _____

18. 2 times 9 are _____

19. 10 times 8 are _____

20. 6 times 3 are _____

21. 2 times 3 times 9

_____ times 9 are _____

22. 2 times 2 times 8

_____ times 8 are _____

23. 3 times 3 times 5

_____ times 5 are _____

24. 4 times 2 times 9

_____ times 9 are _____

25. 2 times 5 times 4

_____ times 4 are _____

Answer five oral questions from Lesson XLVIII, page 51. These questions should be asked by a teacher/tutor.

19 (nineteen)

Add, Subtract, and Multiply

> I can add numbers, subtract numbers, and then multiply numbers.

READ

In Lesson L on page 54, we are reviewing all the things we have learned. We are adding, then subtracting, and then multiplying.

Notice the word **and** tells us to add.

The word **less** tells us to subtract. The word **multiplied by** tells us plainly to multiply. We can also use the word **times**.

STUDY

How many are 2 and 5, less 4, multiplied by 3?

2 and 5 are 7
7 less 4 are 3
3 times 3 are 9

The answer is 9.

RECITE

Complete problems 1 through 13 on this page.

1. 3 and 5 are _____ less 4 are _____ times 2 are _____

2. 4 and 6 are _____ less 5 are _____ times 10 are _____

3. 5 and 5 are _____ less 4 are _____ times 7 are _____

4. 6 and 5 are _____ less 4 are _____ times 7 are _____

5. 7 and 6 are _____ less 5 are _____ times 4 are _____

6. 8 and 7 are _____ less 6 are _____ times 3 are _____

7. 2 and 6 are _____ less 5 are _____ times 4 are _____

8. 3 and 6 are _____ less 5 are _____ times 7 are _____

9. 4 and 7 are _____ less 6 are _____ times 5 are _____

10. 5 and 7 are _____ less 6 are _____ times 8 are _____

11. 6 and 6 are _____ less 5 are _____ times 2 are _____

12. 7 and 7 are _____ less 6 are _____ times 8 are _____

Add, Subtract, and Multiply

I can add numbers, subtract numbers, and then multiply numbers.

READ

In some problems, we have to do two or more things to work the problem. In these problems, we do three things: add, subtract, and then multiply.

Remember that the word **and** means to add, the word **less** means to subtract, and the word **times** means to multiply.

STUDY

How many are 8 and 8, less 7, multiplied by 9?

8 and 8 are 16
16 less 7 are 9
9 times 9 are 81

The answer is 81.

RECITE

Complete problems 15 through 27 on this page.

1. 2 and 7 are _____ less 6 are _____ times 5 are _____

2. 4 and 8 are _____ less 7 are _____ times 9 are _____

3. 3 and 7 are _____ less 6 are _____ times 5 are _____

4. 5 and 8 are _____ less 7 are _____ times 9 are _____

5. 2 and 8 are _____ less 7 are _____ times 6 are _____

6. 3 and 9 are _____ less 8 are _____ times 6 are _____

7. 7 and 10 are _____ less 9 are _____ times 2 are _____

8. 2 and 9 are _____ less 8 are _____ times 7 are _____

9. 3 and 8 are _____ less 7 are _____ times 9 are _____

10. 4 and 10 are _____ less 9 are _____ times 8 are _____

11. 6 and 10 are _____ less 9 are _____ times 8 are _____

12. 6 and 8 are _____ less 7 are _____ times 9 are _____

13. 4 and 9 are _____ less 8 are _____ times 6 are _____

21 (twenty-one)

Story Problems

> **I can use addition, subtraction, and multiplication to work problems.**

READ

In Lesson LI, page 55, we have problems in story form. When we have story problems, we must read the problem carefully. Then we write the problem using numbers. Now we are ready to work the problem and find the answer.

STUDY

Joseph had 14 cents.
He bought 2 oranges at 5 cents each.
How much money had he left?

First find how much the oranges cost.   2 times 5 are 10.
Then find how much money he had left.   14 less 10 are 4.

He had 4 cents left.

RECITE

Use objects if necessary to answer all questions 1 through 10 in Lesson LI, page 55. Ask your teacher/tutor to ask **you** the questions.

Record your answers here.

1. _____4¢_____   6. _____

2. _____   7. _____

3. _____   8. _____

4. _____   9. _____

5. _____   10. _____

22 (twenty-two)

More About Groups

| I can divide an amount into groups. |

READ

We can look at a large group of objects and divide it up into smaller groups. In Lesson LII, page 56, we see the picture of twenty ducks. If we divided the twenty ducks equally into four flocks, how many ducks would be in each flock?

STUDY

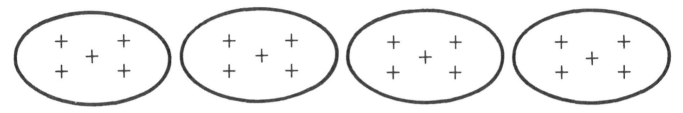

Four flocks of twenty ducks has five ducks in each flock.

RECITE

Draw pictures or use objects if necessary to answer these questions.

1. How many times is 5 contained in 20? _____

2. If the 5 wild geese have 10 wings, how many wings has each

goose? _____

3. How many times is 5 contained in 10? _____

4. Four full-bloomed flowers on 2 branches.

How many on each branch if equally divided? _____

5. How many times is 2 contained in 4? _____

23 (twenty-three)

Review IV

Some problems are worked by doing both addition and subtraction. In this week we worked problems using addition, subtraction, and multiplication.

The word **and** tells us to add. The word **less** tells us to subtract. The word **times** tells us to multiply.

Story problems are written in words. We first must read the words very carefully. Next we write the problem in numbers. Finally we can work the problem. We may need to work more than one problem in some story problems.

When we work with multiplication, we work with groups. When we multiply we combine everything into one big group. We can take the big group and divide it into smaller groups.

Write the numbers on the lines.

1. 3 and 6 are _____ less 5 are _____ times 7 are _____

2. 7 and 7 are _____ less 4 are _____ times 3 are _____

3. 5 and 6 are _____ less 5 are _____ times 4 are _____

4. 3 and 9 are _____ less 8 are _____ times 6 are _____

5. 6 and 10 are _____ less 9 are _____ times 5 are _____

Draw pictures or use objects if necessary to answer these questions.

6. How many times is 5 contained in 20? _____

7. How many times is 2 contained in 10? _____

24 (twenty-four)

Quiz IV

Write the numbers on the lines.

1. 3 and 6 are _____ less 4 are _____ times 3 are _____

2. 8 and 7 are _____ less 6 are _____ times 2 are _____

3. 7 and 4 are _____ less 5 are _____ times 6 are _____

4. 6 and 6 are _____ less 7 are _____ times 4 are _____

5. 3 and 9 are _____ less 4 are _____ times 3 are _____

6. 5 and 8 are _____ less 7 are _____ times 7 are _____

7. 4 and 10 are _____ less 8 are _____ times 3 are _____

8. 9 and 4 are _____ less 6 are _____ times 2 are _____

Draw pictures or use objects to answer this question.

9. How many times is 6 contained in 18?

Answer five oral questions from Lesson LI, page 55. These questions should be asked by a teacher/tutor.

25 (twenty-five)

Division

I can learn division by two.

READ

When we do division, we are still working with groups. Division is working in the opposite direction of multiplication.

When we multiply we have a number of groups with so many in each group. We could have five groups with two in each group. We would multiply and say that five times two are ten.

To divide, we begin with the big number **ten**. We ask how many groups of two are in **ten**. We work backwards and say that we can circle five groups of **two** in ten.

STUDY

How many groups of 2 in 10?

Circle each two objects.

Count the circles or groups.   5

5 groups of 2 are in 10.

RECITE
Write the numbers on the lines.

1. 2 in 2, _____ time
2. 2 in 4, _____ times
3. 2 in 6, _____ times
4. 2 in 8, _____ times
5. 2 in 10, _____ times
6. 2 in 12, _____ times
7. 2 in 14, _____ times
8. 2 in 16, _____ times
9. 2 in 18, _____ times
10. 2 in 20, _____ times

11. 1 in 2, _____ times
12. 2 in 4, _____ times
13. 3 in 6, _____ times
14. 4 in 8, _____ times
15. 5 in 10, _____ times
16. 6 in 12, _____ times
17. 7 in 14, _____ times
18. 8 in 16, _____ times
19. 9 in 18, _____ times
20. 10 in 20, _____ times

Division

| I can use division to work problems. |

READ
Remember that when we do division, we are working with groups. We want to know how many smaller groups are in a large group or how many objects would be in each of the smaller groups.

Begin by using the large number of objects. Divide the objects into groups of the number given.

STUDY
How many apples at 2 cents each can you buy for 4 cents?

2¢

You can buy two apples.

RECITE
Use objects if necessary to answer all questions 1 through 9 in Lesson LIII, page 57. Ask your teacher/tutor to ask **you** the questions.

Record your answers here.

1. _____2 apples_____

2. _____

3. _____

4. _____

5. _____

6. _____

7. _____

8. _____

9. _____

Division

I can learn division by three.

READ

Remember that when we do division, we are still working with groups. Division is working in the opposite direction of multiplication.

Multiplication is finding four groups of three or four groups with three in each group. The total would be four times three are twelve.

To divide, we begin with the big number **twelve**. We work backwards and say that we can circle four groups of **three** in twelve.

STUDY

How many groups of 3 in 12?

Circle each three objects.

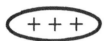 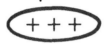  

Count the circles or groups.   4

4 groups of 3 are in 12.

RECITE
Write the numbers on the lines.

1. 3 in 3, _____ time
2. 3 in 6, _____ times
3. 3 in 9, _____ times
4. 3 in 12, _____ times
5. 3 in 15, _____ times
6. 3 in 18, _____ times
7. 3 in 21, _____ times
8. 3 in 24, _____ times
9. 3 in 27, _____ times
10. 3 in 30, _____ times

11. 1 in 3, _____ times
12. 2 in 6, _____ times
13. 3 in 9, _____ times
14. 4 in 12, _____ times
15. 5 in 15, _____ times
16. 6 in 18, _____ times
17. 7 in 21, _____ times
18. 8 in 24, _____ times
19. 9 in 27, _____ times
20. 10 in 30, _____ times

Division

| I can use division to work problems. |

READ

We are working with groups when we do division. We are dividing the large group into smaller groups with the same number in each group. We want to know how many smaller groups or how many objects would be in each of the smaller groups.

Begin by using the large number of objects. Divide the objects into groups of the number given.

STUDY

If you have 6 balls,
how many groups of 3 balls each
can you make out of them?

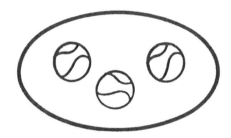 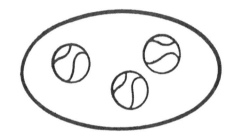

Circle every three balls.

You have 2 groups of 3 balls each in 6 balls.

RECITE

Use objects if necessary to answer all questions 1 through 9 in **Lesson LIV, page 58.** Ask your teacher/tutor to ask **you** the questions.

Record your answers here.

1. _____2 groups_____

2. _____ 6. _____

3. _____ 7. _____

4. _____ 8. _____

5. _____ 9. _____

Review V

When we multiply or divide, we work with groups. Multiplication is finding the big number when we combine one or more groups of some number.

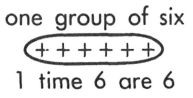

1 time 6 are 6

6 times 1 are 6

Division is beginning with the big number. We divide the big number into groups. We can divide the big number into groups of two. We can ask, "How many groups of 2 in 12?" We can also divide the big number into groups of three. We can ask, "How many groups of 3 in 12?"

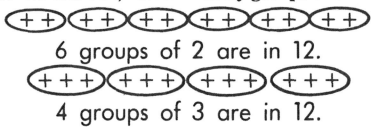

6 groups of 2 are in 12.

4 groups of 3 are in 12.

Write the numbers on the lines.

1. 2 in 4, _____ times
2. 2 in 6, _____ times
3. 2 in 8, _____ times
4. 2 in 10, _____ times
5. 2 in 12, _____ times
6. 2 in 14, _____ times
7. 2 in 16, _____ times
8. 2 in 18, _____ times
9. 2 in 20, _____ times
10. 3 in 6, _____ times
11. 3 in 9, _____ times
12. 3 in 12, _____ times
13. 3 in 15, _____ times
14. 3 in 18, _____ times
15. 3 in 21, _____ times
16. 3 in 24, _____ times
17. 3 in 27, _____ times
18. 3 in 30, _____ times
19. 5 in 10, _____ times
20. 6 in 12, _____ times
21. 7 in 14, _____ times
22. 8 in 16, _____ times
23. 9 in 18, _____ times
24. 10 in 20, _____ times
25. 5 in 15, _____ times
26. 6 in 18, _____ times
27. 7 in 21, _____ times
28. 8 in 24, _____ times
29. 9 in 27, _____ times
30. 10 in 30, _____ times

Quiz V

Write the numbers on the lines.

1. 6 in 12, _____ times

2. 3 in 6, _____ times

3. 4 in 8, _____ times

4. 2 in 14, _____ times

5. 3 in 9, _____ times

6. 2 in 16, _____ times

7. 3 in 15, _____ times

8. 2 in 18, _____ times

9. 2 in 4, _____ times

10. 3 in 27, _____ times

11. 3 in 12, _____ times

12. 2 in 8, _____ times

13. 3 in 30, _____ times

14. 3 in 24, _____ times

15. 2 in 6, _____ times

16. 2 in 20, _____ times

17. 3 in 18, _____ times

18. 5 in 15, _____ times

19. 5 in 10, _____ times

20. 3 in 21, _____ times

21. I can buy _____ pencils at 5 cents each if I have 15 cents.

22. There are _____ in each circle if I divide 12 into 2 groups.

23. I can buy _____ apples at 7 cents each if I have 21 cents.

24. There are _____ in each circle if I divide 18 into 2 groups.

25. There are _____ in each circle if I divide 9 into 3 groups.

Answer five oral questions from Lessons LIII or LIV, pages 57 or 58. These questions should be asked by a teacher/tutor.

31 (thirty-one)

Division

> I can learn division by four.

READ

When we do division, we are still working with groups. Remember that division is working in the opposite direction of multiplication.

Multiplication is finding two groups of four or two groups with four in each group. The total would be two times four are eight.

To divide, we begin with the big number **eight**. We work backwards and say that we can circle two groups of **four** in eight.

STUDY

How many groups of 4 in 8?

Circle each four objects.

Count the circles or groups.   2

2 groups of 4 are in 8.

RECITE
Write the numbers on the lines.

1. 4 in 4, _____ time
2. 4 in 8, _____ times
3. 4 in 12, _____ times
4. 4 in 16, _____ times
5. 4 in 20, _____ times
6. 4 in 24, _____ times
7. 4 in 28, _____ times
8. 4 in 32, _____ times
9. 4 in 36, _____ times
10. 4 in 40, _____ times

11. 1 in 4, _____ times
12. 2 in 8, _____ times
13. 3 in 12, _____ times
14. 4 in 16, _____ times
15. 5 in 20, _____ times
16. 6 in 24, _____ times
17. 7 in 28, _____ times
18. 8 in 32, _____ times
19. 9 in 36, _____ times
20. 10 in 40, _____ times

Division

| I can use division to work problems. |

READ

Division is working with groups. We are dividing the large group into smaller groups with the same number in each group. We want to know how many smaller groups or how many objects would be in each of the smaller groups.

Begin by using the large number of objects. Divide the objects into groups of the number given.

STUDY

How many oranges at 4 cents each can you buy for 8 cents?

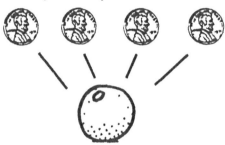
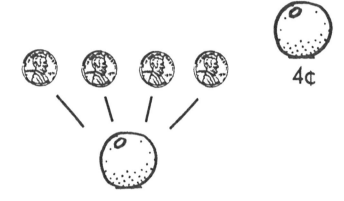

You can buy two oranges.

RECITE

Use objects if necessary to answer all questions 1 through 9 in Lesson LV, page 59. Ask your teacher/tutor to ask **you** the questions.

Record your answers here.

1. _____2 oranges_____

2. _____   6. _____

3. _____   7. _____

4. _____   8. _____

5. _____   9. _____

33 (thirty-three)

Division

I can learn division by five.

READ

Division is working with groups. Division is working in the opposite direction of multiplication.

Multiplication is finding three groups of five or three groups with five in each group. The total would be three times five are fifteen.

To divide, we begin with the big number **fifteen**. We work backwards and say that we can circle three groups of **five** in fifteen.

STUDY

How many groups of 5 in 15?

Circle each five objects.

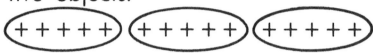

Count the circles or groups.   3

3 groups of 5 are in 15.

RECITE
Write the numbers on the lines.

1. 5 in 5, _____ time
2. 5 in 10, _____ times
3. 5 in 15, _____ times
4. 5 in 20, _____ times
5. 5 in 25, _____ times
6. 5 in 30, _____ times
7. 5 in 35, _____ times
8. 5 in 40, _____ times
9. 5 in 45, _____ times
10. 5 in 50, _____ times

11. 1 in 5, _____ times
12. 2 in 10, _____ times
13. 3 in 15, _____ times
14. 4 in 20, _____ times
15. 5 in 25, _____ times
16. 6 in 30, _____ times
17. 7 in 35, _____ times
18. 8 in 40, _____ times
19. 9 in 45, _____ times
20. 10 in 50, _____ times

Division

| I can use division to work problems. |

READ

Remember that we are working with groups when we do division. We are dividing the large group into smaller groups with the same number in each group. We want to know how many smaller groups or how many objects would be in each of the smaller groups.

Begin by using the large number of objects. Divide the objects into groups of the number given.

STUDY

How many oranges at 5 cents each can you buy for 10 cents?

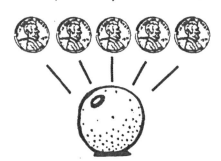 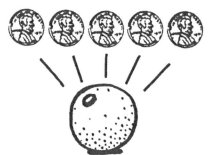 

5¢

You can buy two oranges.

RECITE

Use objects if necessary to answer all questions 1 through 9 in Lesson LVI, page 60. Ask your teacher/tutor to ask you the questions.

Record your answers here.

1. _____2 oranges_____

2. _____   6. _____

3. _____   7. _____

4. _____   8. _____

5. _____   9. _____

35 (thirty-five)

Review VI

In this workbook, we are learning about multiplication and division. When we multiply or divide, we are working with groups. Multiplication is finding the big number when we combine one or more groups of some number.

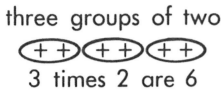

three groups of two
3 times 2 are 6

two groups of three
2 times 3 are 6

Division is beginning with the big number. We divide the big number into groups. We can divide the big number into groups of four. We can ask, "How many groups of 4 in 20?" We can also divide the big number into groups of five. We can ask, "How many groups of 5 in 30?"

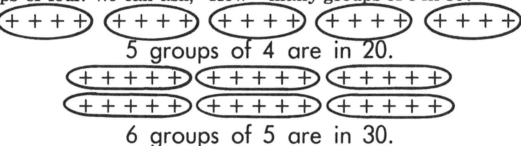

5 groups of 4 are in 20.

6 groups of 5 are in 30.

Write the numbers on the lines.

1. 4 in 8, _____ times
2. 4 in 12, _____ times
3. 4 in 16, _____ times
4. 4 in 20, _____ times
5. 4 in 24, _____ times
6. 4 in 28, _____ times
7. 4 in 32, _____ times
8. 4 in 36, _____ times
9. 4 in 40, _____ times
10. 5 in 10, _____ times
11. 5 in 15, _____ times
12. 5 in 20, _____ times
13. 5 in 25, _____ times
14. 5 in 30, _____ times
15. 5 in 35, _____ times
16. 5 in 40, _____ times
17. 5 in 45, _____ times
18. 5 in 50, _____ times
19. 5 in 20, _____ times
20. 6 in 24, _____ times
21. 7 in 28, _____ times
22. 8 in 32, _____ times
23. 9 in 36, _____ times
24. 10 in 40, _____ times
25. 5 in 25, _____ times
26. 6 in 30, _____ times
27. 7 in 35, _____ times
28. 8 in 40, _____ times
29. 9 in 45, _____ times
30. 10 in 50, _____ times

Quiz VI

Write the numbers on the lines.

1. 4 in 12, _____ times

2. 2 in 10, _____ times

3. 5 in 20, _____ times

4. 9 in 36, _____ times

5. 5 in 15, _____ times

6. 4 in 16, _____ times

7. 5 in 30, _____ times

8. 4 in 36, _____ times

9. 5 in 50, _____ times

10. 4 in 24, _____ times

11. 4 in 32, _____ times

12. 5 in 25, _____ times

13. 8 in 40, _____ times

14. 6 in 30, _____ times

15. 4 in 16, _____ times

16. 9 in 45, _____ times

17. 4 in 20, _____ times

18. 5 in 35, _____ times

19. 10 in 40, _____ times

20. 5 in 10, _____ times

21. There are _____ in each circle if I divide 24 into 4 groups.

22. I can buy _____ oranges at 8 cents each if I have 24 cents.

23. There are _____ in each circle if I divide 35 into 5 groups.

24. I can buy _____ tops at 10 cents each if I have 50 cents.

25. I can buy _____ ribbons at 5 cents each if I have 30 cents.

Answer five oral questions from Lessons LV or LVI, pages 59 or 60.
These questions should be asked by a teacher/tutor.

37 (thirty-seven)

Division

| I can learn division by six. |

READ

Remember we are working with groups when we do division. Division is working in the opposite direction of multiplication.

Multiplication is finding four groups of six or four groups with six in each group. We multiply and say that four times six are twenty-four.

To divide, we start with the big number **twenty-four**. We work backwards and say that we can circle four groups of **six** in twenty-four.

STUDY

How many groups of 6 in 24?

Circle each six objects.

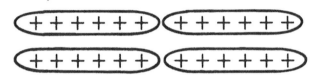

Count the circles or groups.   4

4 groups of 6 are in 24.

RECITE

Write the numbers on the lines.

1. 6 in 6, _____ time

2. 6 in 12, _____ times

3. 6 in 18, _____ times

4. 6 in 24, _____ times

5. 6 in 30, _____ times

6. 6 in 36, _____ times

7. 6 in 42, _____ times

8. 6 in 48, _____ times

9. 6 in 54, _____ times

10. 6 in 60, _____ times

11. 1 in 6, _____ times

12. 2 in 12, _____ times

13. 3 in 18, _____ times

14. 4 in 24, _____ times

15. 5 in 30, _____ times

16. 6 in 36, _____ times

17. 7 in 42, _____ times

18. 8 in 48, _____ times

19. 9 in 54, _____ times

20. 10 in 60, _____ times

Division

I can use division to work problems.

READ

When we do division, we are working with groups. We are dividing the large group into smaller groups with the same number in each group. We want to know how many smaller groups or how many objects would be in each of the smaller groups.

Begin by using the large number of objects. Divide the objects into groups of the number given.

STUDY

How many quarts of milk
at 6 cents a quart
can you buy for 12 cents?

You can buy 2 quarts.

RECITE

Use objects if necessary to answer all questions 1 through 9 in Lesson LVII, page 61. Ask your teacher/tutor to ask **you** the questions.

Record your answers here.

1. _____2 quarts_____

2. _____

3. _____

4. _____

5. _____

6. _____

7. _____

8. _____

9. _____

39 (thirty-nine)

Division

I can learn division by seven.

READ

Remember that when we do division, we are working with groups. Division is working in the opposite direction of multiplication.

Multiplication is finding four groups of seven or four groups with seven in each group. We multiply and say that four times seven are twenty-eight.

To divide, we start with the big number **twenty-eight**. We work backwards and say that we can circle four groups of **seven** in twenty-eight.

STUDY

How many groups of 7 in 28?

Circle each seven objects.

Count the circles or groups.   4

4 groups of 7 are in 28.

RECITE

Write the numbers on the lines.

1. 7 in 7, _____ time
2. 7 in 14, _____ times
3. 7 in 21, _____ times
4. 7 in 28, _____ times
5. 7 in 35, _____ times
6. 7 in 42, _____ times
7. 7 in 49, _____ times
8. 7 in 56, _____ times
9. 7 in 63, _____ times
10. 7 in 70, _____ times

11. 1 in 7, _____ times
12. 2 in 14, _____ times
13. 3 in 21, _____ times
14. 4 in 28, _____ times
15. 5 in 35, _____ times
16. 6 in 42, _____ times
17. 7 in 49, _____ times
18. 8 in 56, _____ times
19. 9 in 63, _____ times
20. 10 in 70, _____ times

Division

| I can use division to work problems. |

READ

We are working with groups when we do division. We are dividing the large group into smaller groups with the same number in each group. We want to know how many smaller groups or how many objects would be in each of the smaller groups.

Begin by using the large number of objects. Divide the objects into groups of the number given.

STUDY

If you divide 14 apples into piles containing 7 apples each, how many piles will there be?

7

7

2 piles of 7 apples each.

RECITE

Use objects if necessary to answer all questions 1 through 9 in Lesson LVIII, page 62. Ask your teacher/tutor to ask **you** the questions.

Record your answers here.

1. _____2 piles_____

2. _____

3. _____

4. _____

5. _____

6. _____

7. _____

8. _____

9. _____

41 (forty-one)

Review VII

Whether we multiply or divide, we are working with groups. Multiplication is finding the big number when we combine one or more groups of some number.

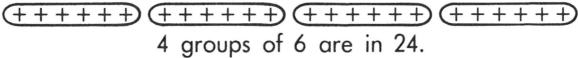

2 times 6 are 12   6 times 2 are 12

Division is beginning with the big number. We divide the big number into groups. We can divide the big number into groups of six. We can ask, "How many groups of 6 in 24?" We can also divide the big number into groups of seven. We can ask, "How many groups of 7 in 21?"

4 groups of 6 are in 24.

3 groups of 7 are in 21.

Write the numbers on the lines.

1. 6 in 12, _____ times
2. 6 in 18, _____ times
3. 6 in 24, _____ times
4. 6 in 30, _____ times
5. 6 in 36, _____ times
6. 6 in 42, _____ times
7. 6 in 48, _____ times
8. 6 in 54, _____ times
9. 6 in 60, _____ times
10. 7 in 14, _____ times
11. 7 in 21, _____ times
12. 7 in 28, _____ times
13. 7 in 35, _____ times
14. 7 in 42, _____ times
15. 7 in 49, _____ times
16. 7 in 56, _____ times
17. 7 in 63, _____ times
18. 7 in 70, _____ times
19. 5 in 30, _____ times
20. 6 in 36, _____ times
21. 7 in 42, _____ times
22. 8 in 48, _____ times
23. 9 in 54, _____ times
24. 10 in 60, _____ times
25. 5 in 35, _____ times
26. 6 in 42, _____ times
27. 7 in 49, _____ times
28. 8 in 56, _____ times
29. 9 in 63, _____ times
30. 10 in 70, _____ times

Quiz VII

Write the numbers on the lines.

1. 6 in 24, _____ times

2. 2 in 12, _____ times

3. 3 in 21, _____ times

4. 6 in 42, _____ times

5. 6 in 54, _____ times

6. 7 in 14, _____ times

7. 4 in 28, _____ times

8. 6 in 18, _____ times

9. 7 in 56, _____ times

10. 5 in 35, _____ times

11. 7 in 63, _____ times

12. 6 in 30, _____ times

13. 9 in 54, _____ times

14. 7 in 42, _____ times

15. 7 in 35, _____ times

16. 6 in 60, _____ times

17. 7 in 49, _____ times

18. 6 in 36, _____ times

19. 9 in 63, _____ times

20. 8 in 48, _____ times

21. I can buy _____ pencils at 9 cents each if I have 54 cents.

22. There are _____ in each circle if I divide 56 into 7 groups.

23. I can buy _____ pears at 7 cents each if I have 35 cents.

24. There are _____ in each circle if I divide 36 into 6 groups.

25. I can buy _____ eggs at 4 cents each if I have 28 cents.

Answer five oral questions from Lessons LVII or LVIII, pages 61 or 62. These questions should be asked by a teacher/tutor.

43 (forty-three)

Division

| I can learn division by eight. |

READ

We are working with groups when we do division. Division is working in the opposite direction of multiplication.

Multiplication is finding three groups of eight or three groups with eight in each group. We multiply and say that three times six are twenty-four.

To divide, we start with the big number **twenty-four**. We work backwards and say that we can circle three groups of **eight** in twenty-four.

STUDY

How many groups of 8 in 24?

Circle each eight objects.

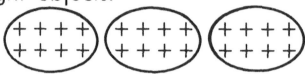

Count the circles or groups.   3

3 groups of 8 are in 24.

RECITE

Write the numbers on the lines.

1. 8 in 8, _____ time

2. 8 in 16, _____ times

3. 8 in 24, _____ times

4. 8 in 32, _____ times

5. 8 in 40, _____ times

6. 8 in 48, _____ times

7. 8 in 56, _____ times

8. 8 in 64, _____ times

9. 8 in 72, _____ times

10. 8 in 80, _____ times

11. 1 in 8, _____ times

12. 2 in 16, _____ times

13. 3 in 24, _____ times

14. 4 in 32, _____ times

15. 5 in 40, _____ times

16. 6 in 48, _____ times

17. 7 in 56, _____ times

18. 8 in 64, _____ times

19. 9 in 72, _____ times

20. 10 in 80, _____ times

Division

I can use division to work problems.

READ

When we do division, we are working with groups. We are dividing the large group into smaller groups with the same number in each group. We want to know how many smaller groups or how many objects would be in each of the smaller groups.

Begin by using the large number of objects. Divide the objects into groups of the number given.

STUDY

If in one peck there are 8 quarts, how many pecks are there in 16 quarts?

8 quarts

8 quarts

8 quarts

2 pecks of 8 quarts each.

RECITE

Use objects if necessary to answer all questions 1 through 9 in Lesson LIX, page 63. Ask your teacher/tutor to ask **you** the questions.

Record your answers here.

1. _____2 pecks_____

2. _____   6. _____

3. _____   7. _____

4. _____   8. _____

5. _____   9. _____

45 (forty-five)

Division

I can learn division by nine.

READ

We are working with groups when we do division. Division is working in the opposite direction of multiplication.

Multiplication is finding two groups of nine or two groups with nine in each group. We multiply and say that two times nine are eighteen.

To divide, we start with the big number **eighteen**. We work backwards and say that we can circle two groups of **nine** in eighteen.

STUDY

How many groups of 9 in 18?

Circle each nine objects.

$$(+ + + + + + + + +)\ (+ + + + + + + + +)$$

Count the circles or groups.   2

2 groups of 9 are in 18.

RECITE

Write the numbers on the lines.

1. 9 in 9, _____ time

2. 9 in 18, _____ times

3. 9 in 27, _____ times

4. 9 in 36, _____ times

5. 9 in 45, _____ times

6. 9 in 54, _____ times

7. 9 in 63, _____ times

8. 9 in 72, _____ times

9. 9 in 81, _____ times

10. 9 in 90, _____ times

11. 1 in 9, _____ times

12. 2 in 18, _____ times

13. 3 in 27, _____ times

14. 4 in 36, _____ times

15. 5 in 45, _____ times

16. 6 in 54, _____ times

17. 7 in 63, _____ times

18. 8 in 72, _____ times

19. 9 in 81, _____ times

20. 10 in 90, _____ times

Division

| I can use division to work problems. |

READ
Remember that when we do division, we are working with groups. We are dividing the large group into smaller groups with the same number in each group. We want to know how many smaller groups or how many objects would be in each of the smaller groups.

Begin by using the large number of objects. Divide the objects into groups of the number given.

STUDY
At 9 cents each, how many pencils can you buy for 18 cents?

9 cents   9 cents   9 cents

2 pencils at 9 cents each.

RECITE
Use objects if necessary to answer all questions 1 through 9 in Lesson LX, page 64. Ask your teacher/tutor to ask **you** the questions.

Record your answers here.

1. _____2 pencils_____

2. _____   6. _____

3. _____   7. _____

4. _____   8. _____

5. _____   9. _____

47 (forty-seven)

Review VIII

Doing multiplication and division cause us to work with groups. Multiplication is finding the big number when we combine one or more groups of some number.

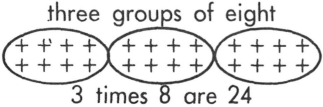

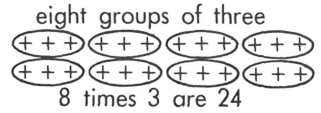

Division is beginning with the big number. We divide the big number into groups. We can divide the big number into groups of eight. We can ask, "How many groups of 8 in 24?" We can also divide the big number into groups of nine. We can ask, "How many groups of 9 in 27?"

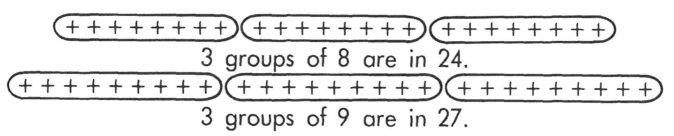

Write the numbers on the lines.

1. 8 in 16, _____ times
2. 8 in 24, _____ times
3. 8 in 32, _____ times
4. 8 in 40, _____ times
5. 8 in 48, _____ times
6. 8 in 56, _____ times
7. 8 in 64, _____ times
8. 8 in 72, _____ times
9. 8 in 80, _____ times
10. 9 in 18, _____ times
11. 9 in 27, _____ times
12. 9 in 36, _____ times
13. 9 in 45, _____ times
14. 9 in 54, _____ times
15. 9 in 63, _____ times
16. 9 in 72, _____ times
17. 9 in 81, _____ times
18. 9 in 90, _____ times
19. 5 in 40, _____ times
20. 6 in 48, _____ times
21. 7 in 56, _____ times
22. 8 in 64, _____ times
23. 9 in 72, _____ times
24. 10 in 80, _____ times
25. 5 in 45, _____ times
26. 6 in 54, _____ times
27. 7 in 63, _____ times
28. 8 in 72, _____ times
29. 9 in 81, _____ times
30. 10 in 90, _____ times

Quiz VIII

Write the numbers on the lines.

1. 2 in 16, _____ times
2. 7 in 56, _____ times
3. 9 in 27, _____ times
4. 8 in 24, _____ times
5. 8 in 72, _____ times
6. 9 in 81, _____ times
7. 8 in 16, _____ times
8. 9 in 27, _____ times
9. 4 in 32, _____ times
10. 9 in 90, _____ times

11. 8 in 48, _____ times
12. 9 in 36, _____ times
13. 6 in 54, _____ times
14. 9 in 18, _____ times
15. 8 in 80, _____ times
16. 5 in 45, _____ times
17. 8 in 40, _____ times
18. 9 in 72, _____ times
19. 8 in 64, _____ times
20. 9 in 63, _____ times

21. There are _____ in each circle if I divide 42 into 7 groups.

22. I can buy _____ apples at 7 cents each if I have 49 cents.

23. There are _____ in each circle if I divide 32 into 4 groups.

24. I can buy _____ tops at 10 cents each if I have 90 cents.

25. There are _____ in each circle if I divide 56 into 8 groups.

Answer five oral questions from Lessons LVIX or LX, pages 63 or 64. These questions should be asked by a teacher/tutor.

49 (forty-nine)

Review I-II

When we multiply, we are working with groups of objects. Each group will have the same number of objects. The group may have one, two, three, four, or five objects in it. The group may have six objects in it.

We may also be talking about how many groups we have. We may have one, two, three, four, or five groups. We may have six groups.

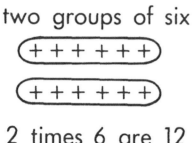

2 times 6 are 12

6 times 2 are 12

The group we are talking about could also have seven, eight, or nine objects in it. Also we could be talking about seven, eight, or nine groups.

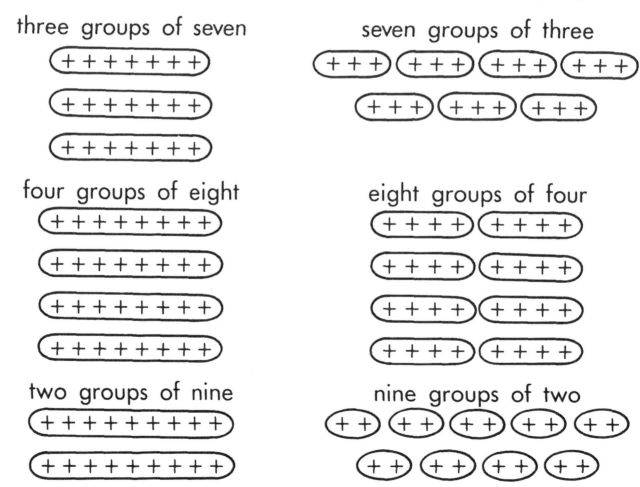

Review pages 47 through 50 in your arithmetic text.

50 (fifty)

Quiz I-II

Write the numbers on the lines.

1. 6 times 2 are _____

2. 7 times 3 are _____

3. 7 times 7 are _____

4. 4 times 7 are _____

5. 8 times 6 are _____

6. 6 times 3 are _____

7. 5 times 6 are _____

8. 8 times 8 are _____

9. 5 times 8 are _____

10. 7 times 5 are _____

11. 4 times 9 are _____

12. 3 times 8 are _____

13. 8 times 9 are _____

14. 8 times 4 are _____

15. 5 times 9 are _____

16. 3 times 9 are _____

17. 6 times 4 are _____

18. 8 times 7 are _____

19. 7 times 6 are _____

20. 9 times 6 are _____

21. 7 nests each having 2 eggs are _____ eggs.

22. 9 boxes each having 7 pencils are _____ pencils.

23. 2 strings of beads with eight beads on a string are _____ beads.

24. Pat has eight dimes. She has _____ cents.

25. Nine apples at 9 cents each costs _____ cents.

Answer ten oral questions from Lessons XLIV through XLVII, pages 47 to 50. These questions should be asked by a teacher/tutor.

51 (fifty-one)

Review III-IV

We have learned to add, subtract, and multiply. When we add, we combine objects. When we subtract, we remove objects. When we multiply, we combine groups of objects.

The word **and** tells us to add. The word **less** tells us to subtract. The word **times** tells us to multiply.

We can combine addition, subtraction, and multiplication in the same problem. We can also have problems where you multiply twice. These problems are two or three problems in one.

Some problems are written in words. We call them story problems. To work a story problem, we must read very carefully. Next we write the problem in numbers. Finally we work the problem.

2 and 4 are __6__ less 2 are __4__ times 7 are __28__
add   subtract   multiply

2 times 3 times 5
multiply   multiply

6 times 5 are 30
multiply

52 (fifty-two)

Quiz III-IV

Write the numbers on the lines.

1. 2 times 10 are _____
2. 10 times 6 are _____
3. 5 times 10 are _____
4. 7 times 10 are _____
5. 10 times 4 are _____

6. 8 times 10 are _____
7. 10 times 9 are _____
8. 3 times 10 are _____
9. 6 times 10 are _____
10. 10 times 10 are _____

11. 3 times 4 are _____
12. 6 times 6 are _____
13. 8 times 7 are _____

14. 3 times 9 are _____
15. 5 times 8 are _____

16. 3 times 3 times 10 _____ times 10 are _____
17. 2 times 4 times 5 _____ times 5 are _____
18. 3 times 2 times 3 _____ times 3 are _____
19. 5 times 2 times 8 _____ times 8 are _____
20. 2 times 3 times 9 _____ times 9 are _____

21. 4 and 7 are _____ less 6 are _____ times 6 are _____
22. 3 and 5 are _____ less 2 are _____ times 4 are _____
23. 2 and 6 are _____ less 5 are _____ times 8 are _____
24. 8 and 7 are _____ less 6 are _____ times 5 are _____
25. 8 and 4 are _____ less 4 are _____ times 4 are _____

Answer ten oral questions from Lessons XLVIII through LI, pages 51 through 55. These questions should be asked by a teacher/tutor.

53 (fifty-three)

Review V-VI

When we divide, we are working with groups of objects. Each group will have the same number of objects. Each group may have one, two, three, four, or five objects in it.

When we divide, we divide the big number of objects into groups. We want to know how many groups we have with the same number of objects in each group.

How many groups of 2 in 8?

Circle every two objects

There are 4 groups of 2 in 8.

How many groups of 3 in 9?

Circle every three objects

There are 3 groups of 3 in 9.

How many groups of 4 in 12?

Circle every four objects

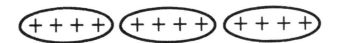

There are 3 groups of 4 in 12.

Quiz V-VI

Write the numbers on the lines.

1. 2 in 12, _____ times

2. 3 in 6, _____ times

3. 4 in 8, _____ times

4. 4 in 32, _____ times

5. 3 in 12, _____ times

6. 4 in 36, _____ times

7. 3 in 9, _____ times

8. 2 in 14, _____ times

9. 3 in 18, _____ times

10. 4 in 16, _____ times

11. 3 in 27, _____ times

12. 5 in 25, _____ times

13. 4 in 12, _____ times

14. 2 in 16, _____ times

15. 5 in 15, _____ times

16. 4 in 24, _____ times

17. 2 in 10, _____ times

18. 5 in 35, _____ times

19. 3 in 24, _____ times

20. 4 in 28, _____ times

21. There are _____ in each circle if I divide 18 into 6 groups.

22. I can buy _____ tops at 5 cents each if I have 20 cents.

23. I can buy _____ apples at 4 cents each if I have 20 cents.

24. There are _____ in each circle if I divide 45 into 5 groups.

25. There are _____ in each circle if I divide 15 into 3 groups.

Answer ten oral questions from Lessons LIII through LVI, pages 57 through 60. These questions should be asked by a teacher/tutor.

55 (fifty-five)

Review VII-VIII

Remember that when we divide, we are working with groups of objects. Each group will have the same number of objects. Each group may have one, two, three, four, five, six, seven, eight, or nine objects in it.

When we divide, we divide the big number of objects into groups. We want to know how many groups we have with the same number of objects in each group.

How many groups of 6 in 24?

Circle every six objects.

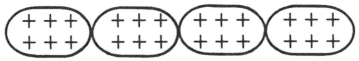

There are 4 groups of 6 in 24.

How many groups of 8 in 24?

Circle every eight objects.

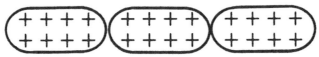

There are 3 groups of 8 in 24.

How many groups of 9 in 27?

Circle every nine objects.

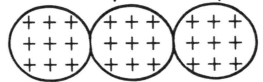

There are 3 groups of 9 in 27.

Quiz VII-VIII

Write the numbers on the lines.

1. 8 in 16, _____ times

2. 6 in 24, _____ times

3. 7 in 56, _____ times

4. 6 in 36, _____ times

5. 8 in 40, _____ times

6. 9 in 18, _____ times

7. 6 in 54, _____ times

8. 8 in 24, _____ times

9. 7 in 49, _____ times

10. 9 in 81, _____ times

11. 9 in 27, _____ times

12. 7 in 28, _____ times

13. 6 in 42, _____ times

14. 8 in 48, _____ times

15. 9 in 36, _____ times

16. 8 in 32, _____ times

17. 8 in 72, _____ times

18. 9 in 63, _____ times

19. 7 in 35, _____ times

20. 8 in 64, _____ times

21. I can buy _____ pencils at 10 cents each if I have 70 cents.

22. I can buy _____ oranges at 8 cents each if I have 40 cents.

23. There are _____ in each circle if I divide 45 into 9 groups.

24. I can buy _____ eggs at 6 cents each if I have 60 cents.

25. There are _____ in each circle if I divide 21 into 7 groups.

Answer ten oral questions from Lessons LVII or LX, pages 61 through 64. These questions should be asked by a teacher/tutor.

57 (fifty-seven)

Ray's New
Primary
Arithmetic
Workbook

Book Three Test

Name _____

Date _____

Possible score _____50_____

My score _____

_____ x _____ = _____%
_{score}

Book Three Test

Write the numbers on the lines.

1. 6 times 4 are _____

2. 9 times 5 are _____

3. 10 times 3 are _____

4. 8 times 7 are _____

5. 4 times 9 are _____

6. 3 times 9 are _____

7. 7 times 6 are _____

8. 4 times 8 are _____

9. 5 times 7 are _____

10. 6 times 3 are _____

11. 8 nests each having 8 eggs are _____ eggs.

12. 9 boxes each having 7 pencils are _____ pencils.

13. 4 strings of beads each having 9 beads are _____ beads.

14. Father has seven dimes. He has _____ cents.

15. Seven apples at 8 cents each costs _____ cents.

16. 3 times 2 times 8 _____ times 8 are _____

17. 4 times 2 times 9 _____ times 9 are _____

18. 2 times 5 times 7 _____ times 7 are _____

19. 3 times 3 times 4 _____ times 4 are _____

20. 2 times 3 times 6 _____ times 6 are _____

21. 6 and 8 are _____ less 5 are _____ times 5 are _____

22. 5 and 7 are _____ less 4 are _____ times 6 are _____

23. 8 and 7 are _____ less 9 are _____ times 4 are _____

24. 9 and 8 are _____ less 8 are _____ times 2 are _____

25. 5 and 8 are _____ less 7 are _____ times 8 are _____

2 (two)

26. 3 in 24, _____ times

27. 8 in 64, _____ times

28. 3 in 18, _____ times

29. 6 in 54, _____ times

30. 5 in 20, _____ times

31. 4 in 24, _____ times

32. 2 in 14, _____ times

33. 7 in 42, _____ times

34. 6 in 36, _____ times

35. 9 in 27, _____ times

36. 9 in 72, _____ times

37. 6 in 48, _____ times

38. 4 in 36, _____ times

39. 3 in 21, _____ times

40. 9 in 54, _____ times

41. 5 in 35, _____ times

42. 4 in 16, _____ times

43. 8 in 48, _____ times

44. 7 in 56, _____ times

45. 8 in 40, _____ times

46. I can buy _____ pencils at 7 cents each if I have 49 cents.

47. There are _____ in each circle if I divide 32 into 4 groups.

48. I can buy _____ apples at 9 cents each if I have 81 cents.

49. There are _____ in each circle if I divide 45 into 9 groups.

50. I can buy _____ tops at 8 cents each if I have 72 cents.

3 (three)

Primary Arithmetic
Series 2 - Book 3
Answer Key

Page 2
1. 6
2. 12
3. 18
4. 24
5. 30
6. 36
7. 42
8. 48
9. 54
10. 60
11. 6
12. 12
13. 18
14. 24
15. 30
16. 36
17. 42
18. 48
19. 54
20. 60

Page 3
1. 12 yards
2. 18 chickens
3. 24 panes
4. 30 peaches
5. 36 cents
6. 42 cents
7. 48 ounces
8. 54 cents
9. 60 cents

Page 3
1. 7
2. 14
3. 21
4. 28
5. 35
6. 42
7. 49
8. 56
9. 63
10. 70
11. 7
12. 14
13. 21
14. 28
15. 35
16. 42
17. 49
18. 56
19. 63
20. 70

Page 5
1. 14 cents,
2. 21 marbles
3. 28 days
4. 35 peaches
5. 42 miles
6. 49 boys
7. 56 marbles
8. 63 cents
9. 70 cents

Page 6
1. 12
2. 18
3. 24
4. 30
5. 36
6. 42
7. 48
8. 54
9. 60
10. 14
11. 21
12. 28
13. 35
14. 42
15. 49
16. 56
17. 63
18. 70
19. 24
20. 30
21. 36
22. 42
23. 48
24. 54
25. 28
26. 35
27. 42
28. 49
29. 56
30. 63

Page 8
1. 8
2. 16
3. 24
4. 32
5. 40
6. 48
7. 56
8. 64
9. 72
10. 80
11. 8
12. 16
13. 24
14. 32
15. 40
16. 48
17. 56
18. 64
19. 72
20. 80

Page 9
1. 16 cents
2. 24 fishes
3. 32 chickens
4. 40 windows
5. 48 pints
6. 56 cents
7. 64 quarts
8. 72 cents
9. 80 cents

Page 10
1. 9
2. 18
3. 27

4. 36
5. 45
6. 54
7. 63
8. 72
9. 81
10. 90
11. 9
12. 18
13. 27
14. 36
15. 45
16. 54
17. 63
18. 72
19. 81
20. 90

Page 11
1. 18 cents
2. 27 cents
3. 36 panes
4. 45 cents
5. 54 cents
6. 63 cents
7. 72 miles
8. 81 cents
9. 90 cents

Page 12
1. 16
2. 24
3. 32
4. 40
5. 48
6. 56
7. 64
8. 72
9. 80
10. 18
11. 27
12. 36
13. 45
14. 54
15. 63
16. 72
17. 81
18. 90
19. 32
20. 40
21. 48
22. 56
23. 64
24. 72
25. 36
26. 45
27. 54
28. 63
29. 72
30. 81

Page 14
1. 10
2. 20
3. 30
4. 40
5. 50
6. 60
7. 70

8. 80
9. 90
10. 100
11. 10
12. 20
13. 30
14. 40
15. 50
16. 60
17. 70
18. 80
19. 90
20. 100

Page 15
1. 20¢
2. $30
3. 40 pecks
4. 50 pounds
5. $60
6. 70 marbles
7. 80¢
8. $90
9. $100

Page 16
1. 12
2. 18
3. 20
4. 9
5. 8
6. 20
7. 16
8. 4
9. 30
10. 30
11. 12
12. 24
13. 40
14. 21
15. 28
16. 25
17. 15
18. 18
19. 36
20. 24
21. 27
22. 32
23. 16
24. 45
25. 35
26. 80
27. 48
28. 70
29. 36
30. 64

Page 17
1. 8,16
2. 6,24
3. 4,20
4. 10,30
5. 6,36
6. 4,24
7. 6,42
8. 4,28
9. 8,64
10. 9,36

1(one)

Answer Key (continued)

Page 18
1. 10
2. 20
3. 30
4. 40
5. 50
6. 60
7. 70
8. 80
9. 90
10. 100
11. 63
12. 40
13. 50
14. 42
15. 54
16. 49
17. 60
18. 56
19. 72
20. 81
21. 8, 40
22. 8, 32
23. 5, 40
24. 10, 70
25. 9, 72

Page 20
1. 8, 4, 8
2. 10, 5, 50
3. 10, 6, 42
4. 11, 7, 49
5. 13, 8, 32
6. 15, 9, 27
7. 8, 3, 12
8. 9, 4, 28
9. 11, 5, 25
10. 12, 6, 48
11. 12, 7, 14
12. 14, 8 64

Page 21
1. 9, 3, 15
2. 12, 5, 45
3. 10, 4, 20
4. 13, 6, 54
5. 10, 3, 18
6. 12, 4, 24
7. 17, 8, 16
8. 11, 3, 21
9. 11, 4, 36
10. 14, 5, 40
11. 16, 7, 56
12. 14, 7, 63
13. 13, 5, 30

Page 22
1. 4¢
2. $20
3. 9 cents
4. $14
5. $9
6. 72 bushels
7. $50
8. 15 cents
9. 28 shoes
10. 54 cents

Page 23
1. 4
2. 2
3. 2
4. 2
5. 2

Page 24
1. 9, 4, 28
2. 14, 10, 30
3. 11, 6, 24
4. 12, 4, 24
5. 16, 7, 35
6. 4
7. 5

Page 26
1. 1
2. 2
3. 3
4. 4
5. 5
6. 6
7. 7
8. 8
9. 9
10. 10
11. 2
12. 2
13. 2
14. 2
15. 2
16. 2
17. 2
18. 2
19. 2
20. 2

Page 27
1. 2 apples
2. 3 marbles
3. 2 lemons
4. 2 peaches
5. 6 yards
6. 2 oranges
7. 8 tops
8. 2 kites
9. 2 books

Page 28
1. 1
2. 2
3. 3
4. 4
5. 5
6. 6
7. 7
8. 8
9. 9
10. 10
11. 3
12. 3
13. 3
14. 3
15. 3
16. 3
17. 3
18. 3
19. 3
20. 3

Page 29
1. 2 groups
2. 3 yards
3. 3 pears
4. 5 yards
5. 6 oranges
6. 3 groups
7. 3 yards
8. 3 cents
9. 10 postage stamps

Page 30
1. 2
2. 3
3. 4
4. 5
5. 6
6. 7
7. 8
8. 9
9. 10
10. 2
11. 3
12. 4
13. 5
14. 6
15. 7
16. 8
17. 9
18. 10
19. 2
20. 2
21. 2
22. 2
23. 2
24. 2
25. 3
26. 3
27. 3
28. 3
29. 3
30. 3

Page 32
1. 1
2. 2
3. 3
4. 4
5. 5
6. 6
7. 7
8. 8
9. 9
10. 10
11. 4
12. 4
13. 4
14. 4
15. 4
16. 4
17. 4
18. 4
19. 4
20. 4

Page 33
1. 2 oranges
2. 3 gallons
3. 4 apples
4. 5 scholars

Page 29 (continued column)
5. 6 copy-books
6. 4 tops
7. 8 peaches
8. 4 cakes
9. 4 books

Page 34
1. 1
2. 2
3. 3
4. 4
5. 5
6. 6
7. 7
8. 8
9. 9
10. 10
11. 5
12. 5
13. 5
14. 5
15. 5
16. 5
17. 5
18. 5
19. 5
20. 5

Page 35
1. 2 oranges
2. 5 pencils
3. 5 toy books
4. 5 pears
5. 5 melons
6. 5 weeks
7. 5 cakes
8. 9 tops
9. 10 slates

Page 36
1. 2
2. 3
3. 4
4. 5
5. 6
6. 7
7. 8
8. 9
9. 10
10. 2
11. 3
12. 4
13. 5
14. 6
15. 7
16. 8
17. 9
18. 10
19. 4
20. 4
21. 4
22. 4
23. 4
24. 4
25. 5
26. 5
27. 5
28. 5
29. 5
30. 5

2(two)

Answer Key (continued)

Page 38
1. 1
2. 2
3. 3
4. 4
5. 5
6. 6
7. 7
8. 8
9. 9
10. 10
11. 6
12. 6
13. 6
14. 6
15. 6
16. 6
17. 6
18. 6
19. 6
20. 6

Page 39
1. 2 quarts
2. 3 oranges

3. 4 trees
4. 6 pears
5. 6 pounds
6. 6 lemons
7. 6 pencils
8. 9 rings
9. $10

Page 40
1. 1
2. 2
3. 3
4. 4
5. 5
6. 6
7. 7
8. 8
9. 9
10. 10
11. 7
12. 7
13. 7
14. 7
15. 7

16. 7
17. 7
18. 7
19. 7
20. 7

Page 41
1. 2 piles
2. 3 pineapples
3. 4 melons
4. 5 peaches
5. 7 miles
6. 7 cents
7. 7 trees
8. 7 yards
9. $10

Page 42
1. 2
2. 3
3. 4
4. 5
5. 6

6. 7
7. 8
8. 9
9. 10
10. 2
11. 3
12. 4
13. 5
14. 6
15. 7
16. 8
17. 9
18. 10
19. 6
20. 6
21. 6
22. 6
23. 6
24. 6
25. 7
26. 7
27. 7
28. 7
29. 7
30. 7

Primary Arithmetic
Series 2 - Book 3
Quizzes and Tests

Quiz I
1. 12
2. 28
3. 24
4. 18
5. 24
6. 42
7. 49
8. 48
9. 30
10. 21
11. 21
12. 54
13. 35
14. 48
15. 60
16. 27
17. 35
18. 56
19. 63
20. 70
21. 35
22. 50
23. 18
24. 9
25. 56

Quiz II
1. 16
2. 36
3. 32
4. 24
5. 32
6. 54
7. 63

8. 64
9. 40
10. 27
11. 24
12. 72
13. 45
14. 48
15. 80
16. 72
17. 45
18. 18
19. 81
20. 90
21. 90
22. 40
23. 90
24. 32
25. 48

Quiz III
1. 50
2. 8
3. 42
4. 90
5. 30
6. 20
7. 21
8. 36
9. 40
10. 45
11. 56
12. 28
13. 30
14. 20
15. 64

16. 60
17. 15
18. 18
19. 80
20. 18
21. 6, 54
22. 4, 32
23. 9, 45
24. 8, 72
25. 10, 40

Quiz IV
1. 9, 5, 15
2. 15, 9, 18
3. 11, 6, 36
4. 12, 5, 20
5. 12, 8, 24
6. 13, 6, 42
7. 14, 6, 18
8. 13, 7, 14
9. 3

Quiz V
1. 2
2. 2
3. 2
4. 7
5. 3
6. 8
7. 5
8. 9
9. 2
10. 9

11. 4
12. 4
13. 10
14. 8
15. 3
16. 10
17. 6
18. 3
19. 2
20. 7
21. 3
22. 6
23. 3
24. 9
25. 3

Quiz VI
1. 3
2. 5
3. 4
4. 4
5. 3
6. 4
7. 6
8. 9
9. 10
10. 6
11. 8
12. 5
13. 5
14. 5
15. 4
16. 5
17. 5

18. 7
19. 4
20. 2
21. 6
22. 3
23. 7
24. 5
25. 6

Quiz VII
1. 4
2. 6
3. 7
4. 7
5. 9
6. 2
7. 7
8. 3
9. 8
10. 7
11. 9
12. 5
13. 6
14. 6
15. 5
16. 10
17. 7
18. 6
19. 7
20. 6
21. 6
22. 8
23. 5
24. 6
25. 7

3(three)

Answer Key (continued)

Page 44
1. 1
2. 2
3. 3
4. 4
5. 5
6. 6
7. 7
8. 8
9. 9
10. 10
11. 8
12. 8
13. 8
14. 8
15. 8
16. 8
17. 8
18. 8
19. 8
20. 8

Page 45
1. 2 pecks
2. 3 oranges

3. 8 pencils
4. 8 cents
5. 6 canes
6. 8 tops
7. 8 peaches
8. 8 cents
9. 10 cents

Page 46
1. 1
2. 2
3. 3
4. 4
5. 5
6. 6
7. 7
8. 8
9. 9
10. 10
11. 9
12. 9
13. 9
14. 9
15. 9

16. 9
17. 9
18. 9
19. 9
20. 9

Page 47
1. 2 pencils
2. 9 cents
3. 9 cents
4. 9 cents
5. 6 cents
6. 9 miles
7. 8 yards
8. 9 blocks
9. 9 books

Page 48
1. 2
2. 3
3. 4
4. 5
5. 6

6. 7
7. 8
8. 9
9. 10
10. 2
11. 3
12. 4
13. 5
14. 6
15. 7
16. 8
17. 9
18. 10
19. 8
20. 8
21. 8
22. 8
23. 8
24. 8
25. 9
26. 9
27. 9
28. 9
29. 9
30. 9

Quizzes & Tests (continued)

Quiz VIII
1. 8
2. 8
3. 3
4. 3
5. 9
6. 9
7. 2
8. 3
9. 8
10. 10
11. 6
12. 4
13. 9
14. 2
15. 10
16. 9
17. 5
18. 8
19. 8
20. 7
21. 6
22. 7
23. 8
24. 9
25. 7

Quiz I-II
1. 12
2. 21
3. 49
4. 28
5. 48
6. 18
7. 30
8. 64
9. 40

10. 35
11. 36
12. 24
13. 72
14. 32
15. 45
16. 27
17. 24
18. 56
19. 42
20. 54
21. 14
22. 63
23. 16
24. 80
25. 81

Quiz III-IV
1. 20
2. 60
3. 50
4. 70
5. 40
6. 80
7. 90
8. 30
9. 60
10. 100
11. 12
12. 36
13. 56
14. 27
15. 40
16. 9, 90
17. 8, 40
18. 6, 18
19. 10, 80

20. 6, 54
21. 11, 5, 30
22. 8, 6, 24
23. 8, 3, 24
24. 15, 9, 45
25. 12, 8, 32

Quiz V-VI
1. 6
2. 2
3. 2
4. 8
5. 4
6. 9
7. 3
8. 6
9. 4
10. 4
11. 9
12. 5
13. 3
14. 8
15. 3
16. 6
17. 5
18. 7
19. 8
20. 7
21. 3
22. 4
23. 5
24. 9
25. 5

Quiz VII-VIII
1. 2
2. 4
3. 8

4. 6
5. 5
6. 2
7. 9
8. 3
9. 7
10. 9
11. 3
12. 4
13. 7
14. 6
15. 4
16. 4
17. 9
18. 7
19. 5
20. 8
21. 7
22. 5
23. 5
24. 10
25. 3

Book Three Test
1. 24
2. 45
3. 30
4. 56
5. 36
6. 27
7. 42
8. 32
9. 35
10. 18
11. 64
12. 63
13. 36

14. 70
15. 56
16. 6, 48
17. 8, 72
18. 10, 70
19. 9, 36
20. 6, 36
21. 14, 9, 45
22. 12, 8, 48
23. 15, 6, 24
24. 17, 9, 18
25. 13, 6, 48
26. 8
27. 8
28. 6
29. 9
30. 4
31. 6
32. 7
33. 6
34. 6
35. 3
36. 8
37. 8
38. 9
39. 7
40. 6
41. 7
42. 4
43. 6
44. 8
45. 5
46. 7
47. 8
48. 9
49. 5
50. 9